AF882721

O CHEM VOL. II

Place of Peace Press

ISBN:978-1-954514-03-4

For more books like this visit peacefulpress.net

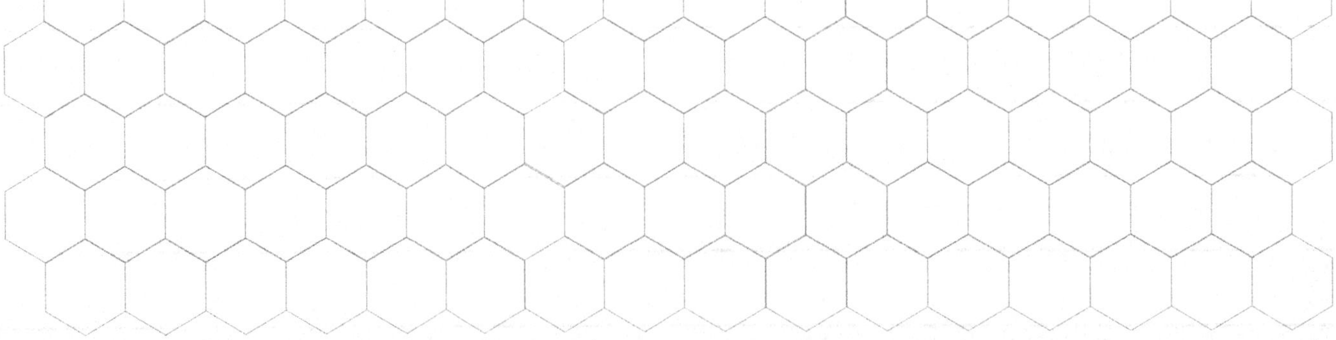

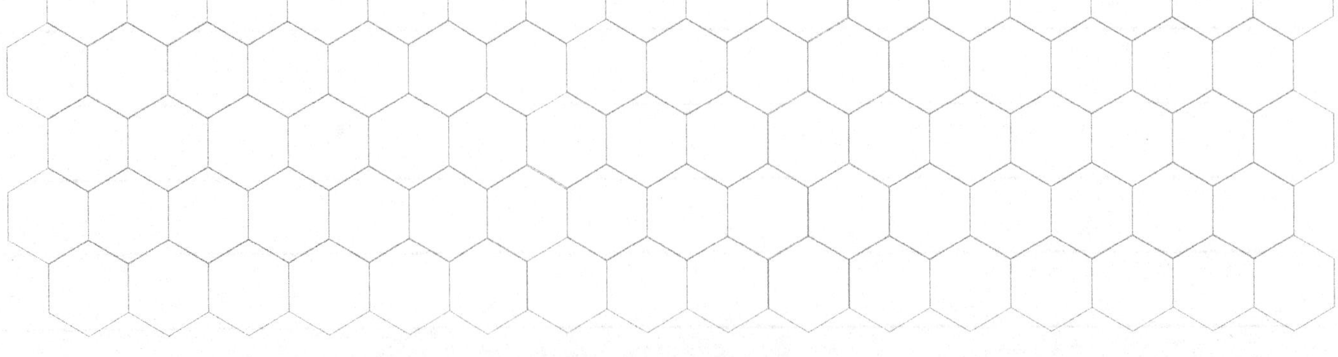

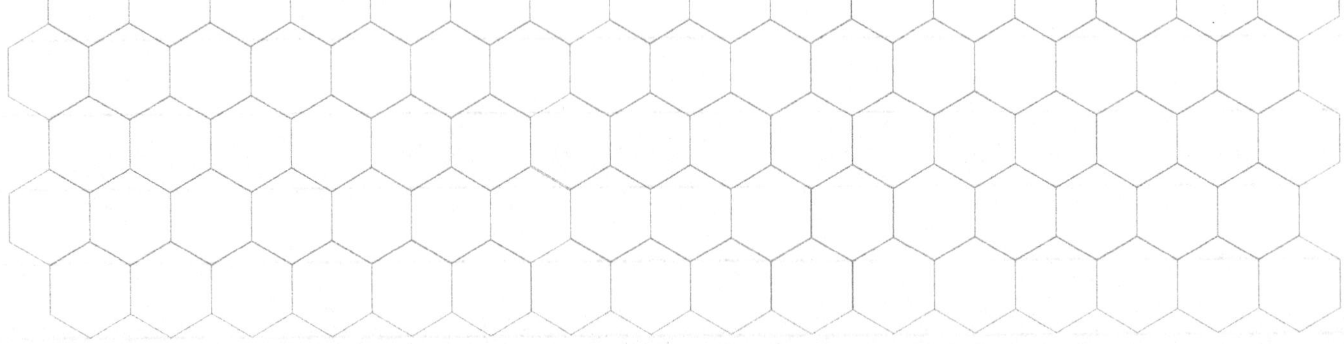

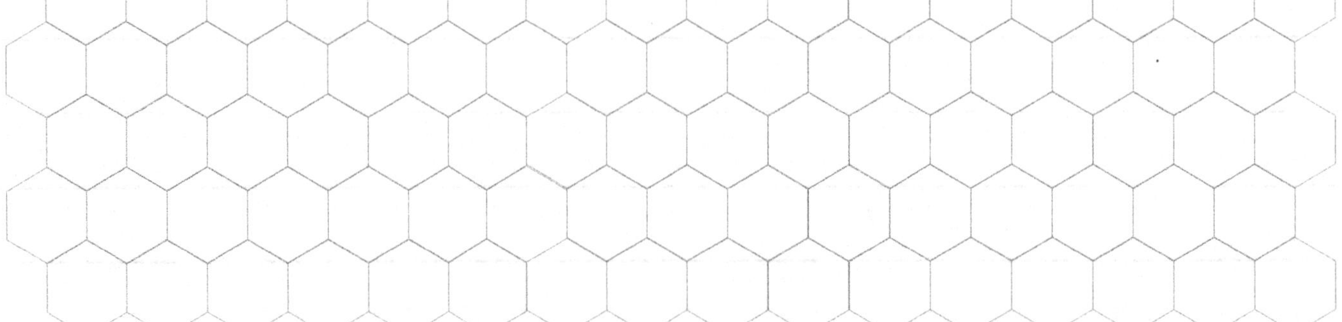

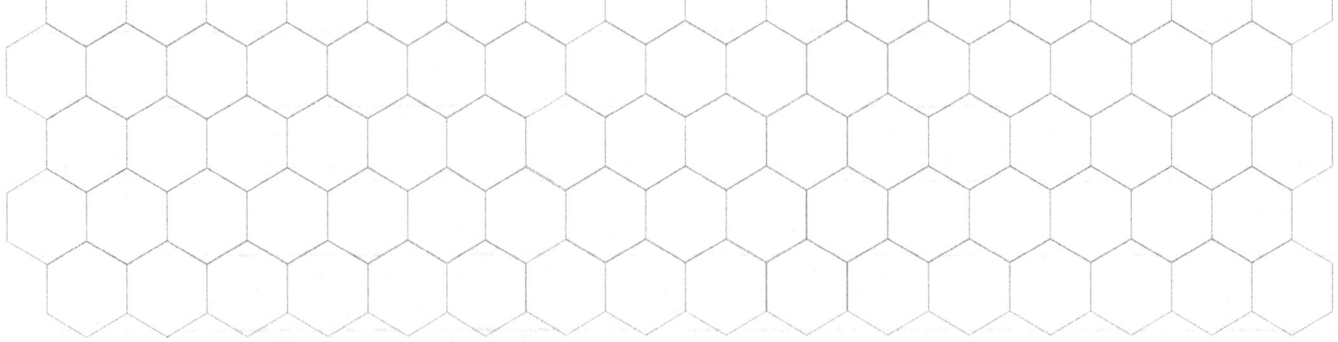

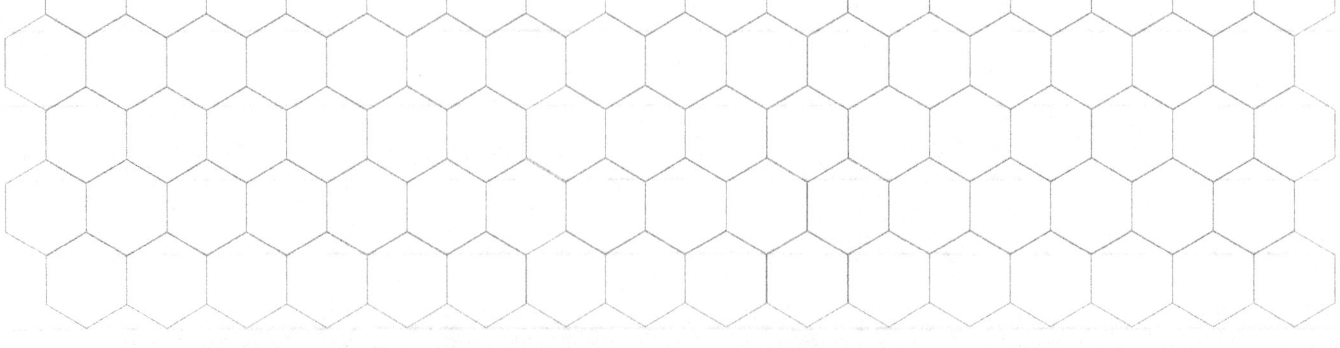

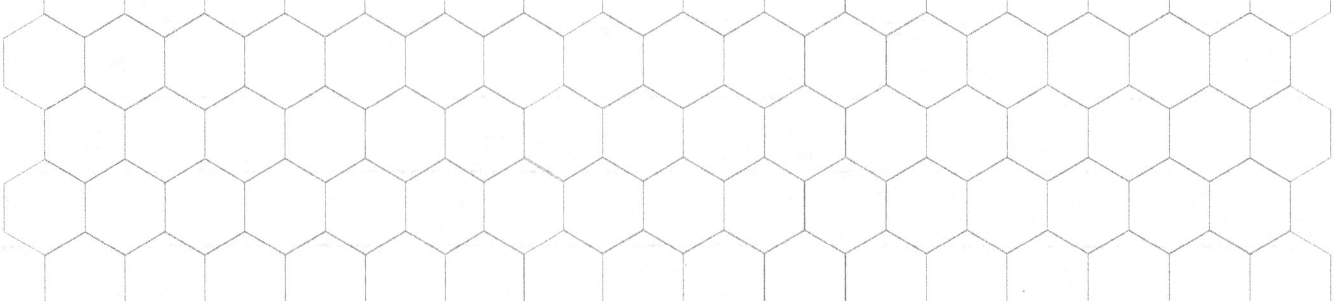

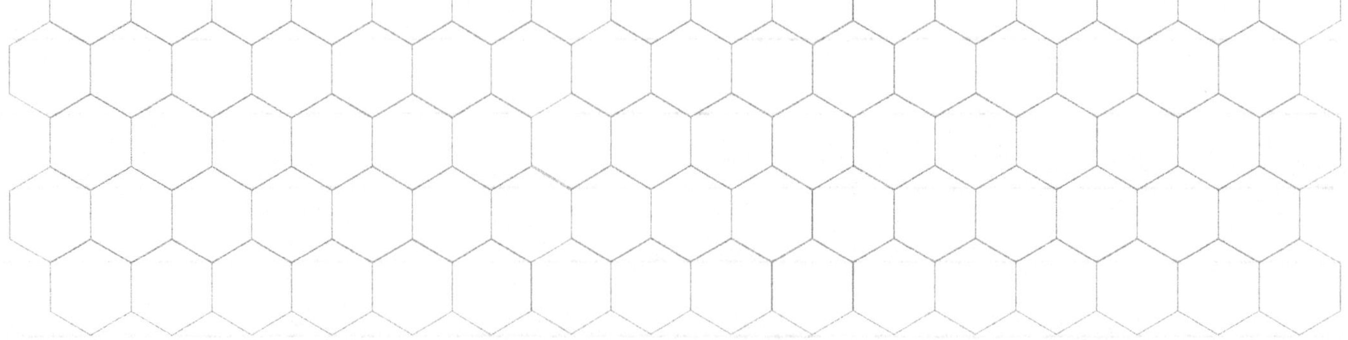

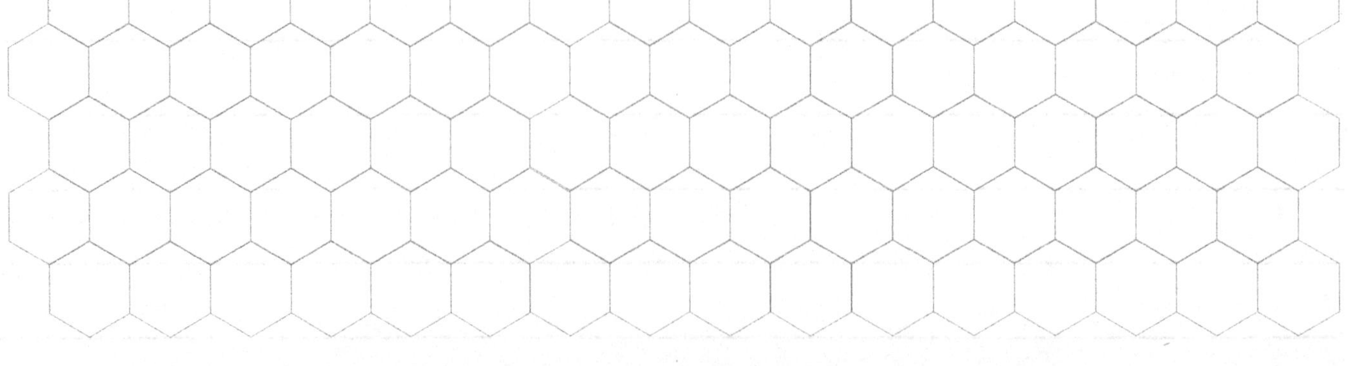

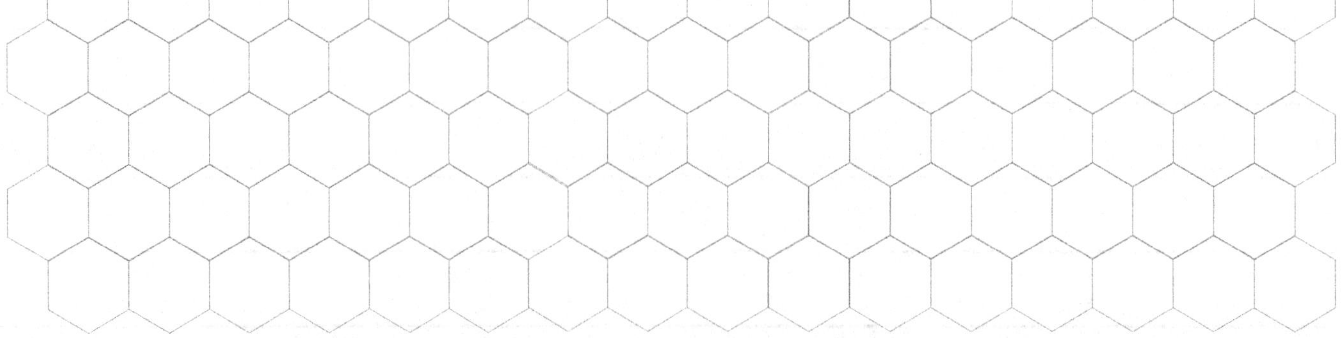

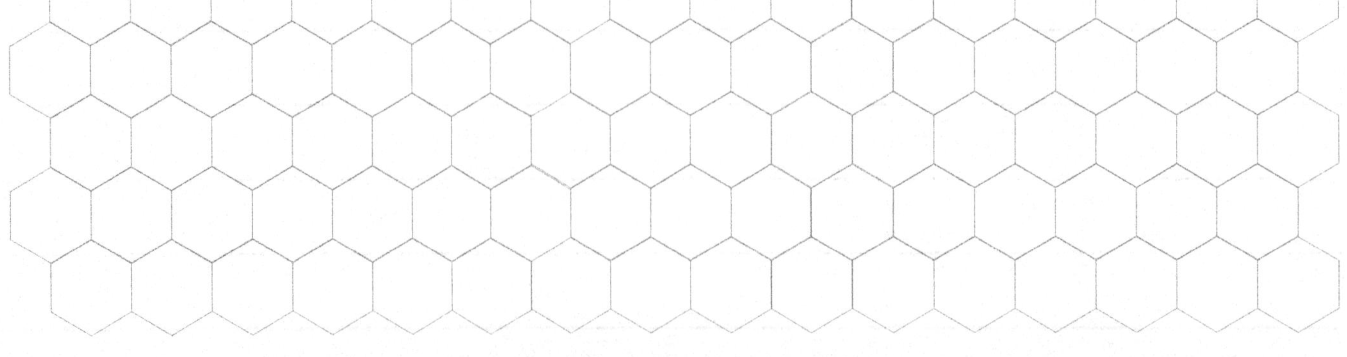

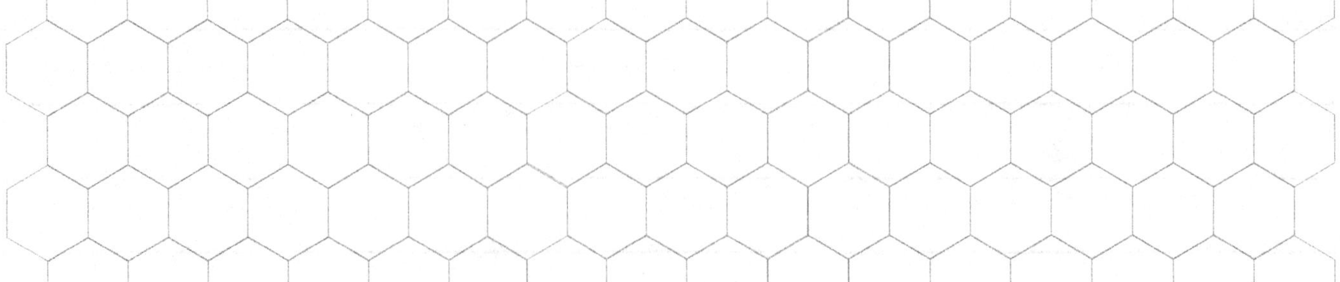

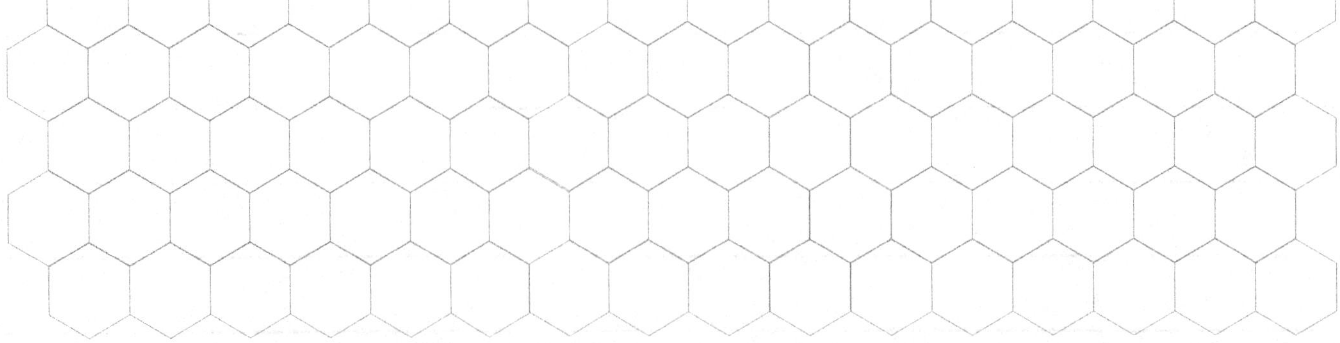

ORGANIC FUNCTIONAL GROUPS

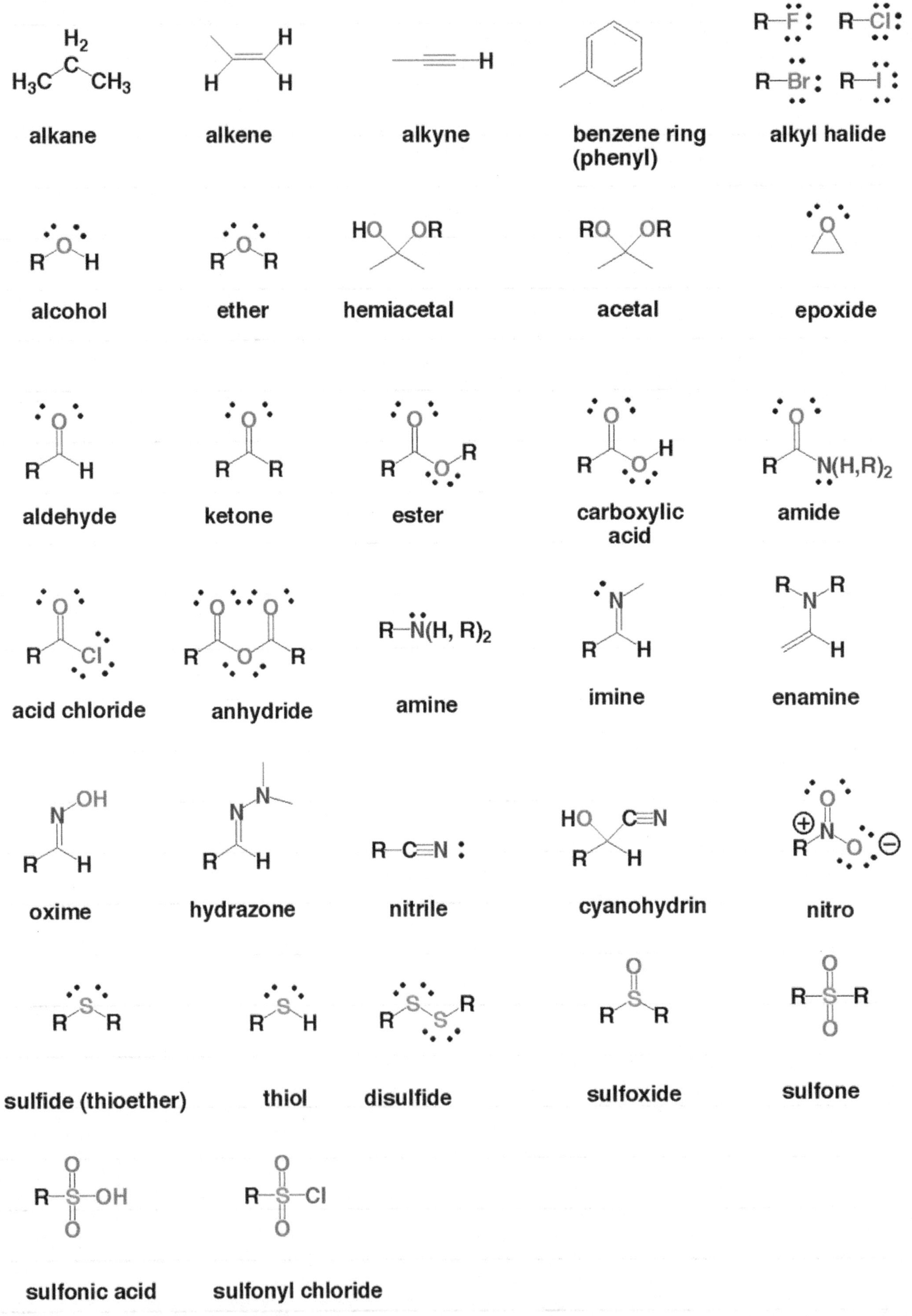

Made in the USA
Monee, IL
08 November 2026

67972899R00083